AF443249

THE WORLD OF
WILSONS PROMONTORY

THE WORLD OF
WILSONS PROMONTORY

JOCELYN BURT

Picnic Point

Norman Beach

The Five Mile Press

First published 1983
The Five Mile Press Pty. Ltd.,
85 Canterbury Road,
Canterbury, Victoria,
Australia, 3126

Designed by Derrick I. Stone Design
Typesetting by Meredith Trade Lino, Melbourne
Printed in Hong Kong by Dai Nippon Printing Company Ltd.

National Library of Australia
Cataloguing-in-Publication Data.

Burt, Jocelyn.
The World of Wilsons Promontory.

ISBN 0 86788 021 X.

1. National parks and reserves — Victoria —
Description and travel — Views. 2. Wilsons'
Promontory National Park (Vic.) —
Description and travel — Views. I. Title.

919.945'60463

For J.M.B.

CONTENTS

ACKNOWLEDGEMENTS

I am indebted to the administration staff and rangers of Wilsons Promontory National Park for their help in the preparation of this book. In particular I wish to thank Peter and Merran Thomas, Roy Speechley, Steve Voros, Jim Whelan, and Gill Anderson.

I would also like to thank Allen and Marlene Levings for their help at the Wilsons Promontory lighthouse.

PREFACE

My love affair with Wilsons Promontory began in 1974, when John Béchervaise introduced me to the area while we were working on our book *Wilsons Promontory* for the Rigby Conservation Series. It didn't take long for me to join the thousands of people caught in its spell. Since then, every visit has brought for me a deeper love and awareness because the Prom — as it is affectionately called by Victorians — has a unique character and beauty. It's hardly surprising that it is Victoria's most popular national park — and amongst the most frequently visited in Australia.

The joy I have had in preparing both this, and the earlier book, has been enormous; certainly I felt much fitter after all the walking! Like all who visit the Prom, I soon discovered it cannot be explored properly in a few visits. Even during the time it took for me to prepare both books, I had the constant frustrating awareness I was touching only the surface. It would take a lifetime to know this place really well. But this is one of the great delights of the Prom: each time you visit, you will *always* find something new and different to enjoy — previously undiscovered beauty and moods, regardless of the season.

In this book I would like to share some of the beauty I have experienced, and to give prospective walkers not only a glimpse of the joys that lie ahead, but also some idea of general walking conditions. For those who know the Prom well, I hope I may be able to stir nostalgic memories of past walks in this magnificent and much-loved place.

Yanakie
Scale:
0 1 2 3 4 5 6 7 8 9 10 Km
Shellback Island
Darby Beach
Tongue Point
Whisky Bay
Picnic Bay
Norman Island
Mt. Bishop
Lilly Pilly Gully
Sealer's Cove
Tidal River
Mt. Oberon
Refuge Cove
Glennie Group
Great Glennie Island
Oberon Bay
Waterloo Bay
The Lighthouse
Cleft I.
Anser I.
Wattle I.
Kanowna Is.
N
Location:
Australia
Wilson's Promontory

WILSONS PROMONTORY

In wild splendour the high granite peninsula of Wilsons Promontory pushes into the cold waters of Bass Strait to form the most southerly point of the Australian mainland, 225 kilometres south-east of Melbourne. In a relatively small area this national park protects an extraordinarily diverse range of environments: wild heathlands and swamps, sweet-scented woodlands, moist pockets of rainforest, granite mountain tops, and dunes and seashores. All support an abundance of interesting flora and fauna. Wilsons Promontory is also a world of exceptional beauty, marked as it is by that wonderful combination of mountain and sea. Its forest-clad slopes drop steeply to rocky shores and beaches of fine sand washed by foaming surf or lapped by the sheltered waters of small coves. Some of the beaches are wide and spacious, others curve gracefully into the coast between prominent headlands, or fit snugly between masses of boulders. On most there is a feeling of wilderness, and all epitomise the essential and precious freedom that Australia offers so generously along its shores.

If the beaches are a major contributor to the beauty of Wilsons Promontory, the granite is its strength and character. Its peaks and slopes are strewn with tors of astonishing proportions, or encrusted with massive slabs that fall away as sheer cliffs. Promontories and lesser headlands are heavily seamed with cracks and studded with boulders of every shape and size. Along the coastline the restless sea heaves its powerful swells and crashes its great, white-capped waves against the hard granite. Many of the beaches have a splendid collection of boulders heaped at one or both ends. Often these grey tors are decorated with orange lichen, and are tumbled in curious positions that fascinate the imagination and invite exploration — access permitting.

More granite lies in the deep gullies of the hills. Here the stones and boulders are shrouded in thick green moss, and shelter moisture-loving plants and streams of clear, sparkling water that gurgle merrily over rocky beds hidden in the undergrowth. Often the smaller boulders are canopied by huge sprays of tree-ferns that flourish in the Promontory's damp corners. There is no escaping the granite — it is everywhere. And how welcome it is, when a weary hiker finds a seat-sized boulder strategically placed beside the walking track.

Walking the Prom is a sheer delight. Over twenty maintained tracks totalling around 130 kilometres lead to most parts of the national park, and one of the great joys of the place is the enormous variety of walks available to visitors of all ages — and all degrees of fitness. There are short and easy ones, and specially designed nature trails; there are half-day and full-day walks, and longer hikes for which outstation camping is provided. In order to prevent overcrowding at the outstation camping grounds, permits are required for all overnight walks, and these may be obtained from the park office at Tidal River. Wood fires are prohibited in some areas because of the scarcity of firewood; they are also banned totally at certain times of the year because of the danger of bushfires.

Over the years, Wilsons Promontory has suffered severely from destructive fires, most started by man — either deliberately, by pastoralists during the last century, or more recently through the carelessness of picnickers, campers and fishermen. The worst fire broke out in the summer of 1951, when nearly three-quarters of the national park was reduced to ashes. Fortunately both the settlement at Tidal River and the hundreds of campers who were there escaped the fire. However, the loss of wildlife was appalling, as the fire occurred in the middle of the nesting season for both resident and migrant birds.

Bushfires have been the cause of the change in vegetation in many parts of the Promontory. When too many fires pass over the same ground, the original plant communities are prevented from re-establishing themselves; instead, hardier species replace them. Areas that once bore forests of tall eucalypts are covered with impenetrable scrub. Today tall, grey skeletons that rear over the vegetation of the mountain slopes are a stark reminder of this enforced evolution, and of its cause.

Visitors sometimes grumble at the Promontory's notoriously bad weather. Yet it is on wild and stormy days that the Prom's essential character is often most dramatically revealed. In calm, sunny weather the scenery is spectacular enough, but when billowing grey and white clouds threaten behind craggy, boulder-strewn peaks and headlands, and the sun slips through to spotlight a gleaming white beach or a group of huge, orange-capped tors outlined against a grey-black sky, Wilsons Promontory really comes alive. Apart from this dramatic beauty, the cooler, blustery days are excellent for walking, and even the delectable, ozone-laden air seems fresher and healthier.

The most-stable weather occurs between January and March, the most popular time with holidaymakers. Winters are surprisingly mild, and cooling breezes generally keep the summer temperature at a pleasantly bearable level — although there can be the odd day of searing heat. Many who know and love the Prom prefer the spring, when wildflowers bloom in splendid profusion, and the capricious spring weather patterns create the constantly changing moods that give the area its most stirring beauty. At this time, too, it is quieter around the camping grounds.

Leonard Bay. On blustery days, when the sun chases sea mists drifting through the air and the dull roar of the ocean dominates all other sounds, it is quite exhilarating to watch the powerful waves unleash their energy in explosions of white foam against the rocks.

EARLY DAYS

The Promontory's granite was formed during the Palaeozoic Era, about 300 million years ago. This grey rock began as a molten mass that cooled slowly, deep within the Earth's crust, and was gradually revealed as the older overlying rock material was eroded away. Originally the entire Promontory was part of the land mass that linked Tasmania to mainland Australia. However, when the polar icecaps melted at the end of the last Ice Age — only about 10,000 years ago — the Bass Strait region was submerged, leaving numerous granite mountain tops as islands above sea level; Wilsons Promontory was in fact one of these islands. Over the years it was re-joined to the mainland by accumulated deposits of blown sand which gradually became stabilised by natural vegetation, forming the low Yanakie isthmus that now separates Corner Inlet to the east, and Waratah Bay to the west.

Long before white men came to these shores, Wilsons Promontory was occupied by the Brataualung tribe of Aborigines, who collected food along the shores during the warmer months. The only evidence of their habitation is seen in their kitchen middens — the heaps of broken shellfish shells lying in the more sheltered dunes of the west coast. Occasionally, drifting sand may expose an axe-head, a stone chip, or some other ancient artifact.

The first European to sight the Promontory was George Bass, who made his discovery on 2 January 1798, during his epic voyage in a nine-metre whaleboat with a crew of six oarsmen. He sailed from Sydney to explore the south-eastern coast, a journey covering over 1900 kilometres. Today one trembles at the thought of such a venture, knowing the waters of Bass Strait can so easily change from comparative calm to mountainous seas whipped by gale force winds. Wilsons Promontory was named after Thomas Wilson, a family friend of Bass and Matthew Flinders.

During the first hundred years of settlement, the Promontory was exploited mercilessly. It is suspected that thousands of Koalas and wallabies were killed for their skins; certainly the region's enormous population of Fur Seals was reduced to alarmingly low levels by hunters. Sealers Cove was a major base for sealing operations. (Since the wholesale slaughtering stopped in the 1830s, seal colonies on some of the islands have built up to moderate numbers.) Another early industry was logging, for which again Sealers Cove was used as a base. The timber felled on the slopes was brought down to the beach by flying fox or wood-railed tramways. It wasn't long before the pastoralists moved in, and small numbers of cattle were grazed around the Sealers Cove area as well as Darby and Oberon bays. Towards the end of the nineteenth century, a settlement was planned for the Corner Inlet area of the Promontory. And even after the national park was declared, tin was mined for a few years at Mount Hunter in the north.

Fortunately, naturalists were quick to realise that the area was a botanic and zoological treasure chest. In 1884 three enthusiastic members of the Field Naturalists Club of Victoria spent two weeks exploring the Prom; shortly afterwards they published an article on their trip in the *Victorian Naturalist*, expressing the hope that Wilsons Promontory would become a haven for nature lovers. It was not until 1887, when the settlement scheme was planned, that the Field Naturalist Club decided to campaign for a national park. It was a long battle, and when eventually it was declared a national park in 1905, none of its shoreline was included — the boundary ended nearly a kilometre away from the coast! The naturalists saw this as a most undesirable situation, and three years later the boundary was widened to embrace the coastal areas. Additions since then, including part of the Yanakie isthmus and many of the off-shore islands, have brought the park to 48,935 hectares, or nearly 300 square kilometres.

Once declared a national park, the Promontory became a Mecca for nature lovers. In 1923 a chalet opened at Darby River, and two years later huts were erected at Tidal River, Corner Inlet and Sealers Cove. During the Second World War the park was closed to the public while the army used it as a commando training camp. Until this time access had been dificult. Before the wartime road to Tidal River was built, many people took the boat from Port Franklin or Port Welshpool to the southern edge of Corner Inlet, and from there walked to the Darby Chalet; some groups travelled on horseback. Others drove to the Chalet from Yanakie, taking the vehicles a good part of the way along Darby Beach. A few intrepid hikers even walked in from Fish Creek, and, after reaching the Chalet, continued on to the lighthouse and other spots at the southern end. Before the walking tracks were established, hikers often followed the beaches and headlands in order to reach many of the bays and coves we visit today. J. Ros. Garnet, author of *Wildflowers of Wilsons Promontory*, made his first trip to the park in 1924 on a bicycle, following the beach along Waratah Bay. Such exploits say a great deal for the earlier generation of visitors, whose tremendous sense of adventure and physical fortitude must be admired.

The road through the Yanakie isthmus to Tidal River was at first a mere sandy track, which was gradually upgraded and finally sealed in the early 1970s. These days Tidal River is a well-run holiday resort with flats, lodges, and huts nestling among the tea-trees and banksias. A maze of tracks leads to camping sites tucked in the surrounding vegetation. There is a general store and a medical centre, and the National Parks Service has designed and built an excellent Information Centre which is one of the most comprehensive in Australia.

With an abundance of magnificent scenery, and the prospect of some good healthy walking in an unpolluted atmosphere, it is small wonder that by the early 1970s the Promontory was receiving around 80,000 visitors a year. A decade later the figure had reached a quarter of a million.

TIDAL RIVER

The Tidal River resort is ideally situated in a corner of the valley lying between Little Oberon and Pillar Point; bordered on one side by the meandering Tidal River, and on the other by the broad beach of Norman Bay. As its name suggests, the river is tidal. Running close to the headland, it carves a new course twice daily through the sands which extend for some distance along its southern side. A few steps from the camping ground take you to these sands — and to a constantly changing scene that is dominated to the east by Mount La Trobe, the Promontory's highest mountain (754 metres).

Take a walk at sunrise, when few — if any — people are around. Once the sun has climbed over the hill, soft light spills through the valley and chases away the shadows that darken the rocks lining the river. If there is no wind to ruffle the surface of the water, the reflections are wonderful. The huge boulders, many with their sober granite-grey splashed with colourful lichen, are mirrored so perfectly that they seem to balance on their own shapes. On such a day, one of the most spectacular sights is Whale Rock, situated near the footbridge. Vaguely resembling the head of a whale, and whale-like in size, this great rock props on an equally massive slab of granite, so that it towers over the other boulders, river, and vegetation.

Each day the scene along Tidal River changes. Sometimes misty white clouds wreath Mount La Trobe, and if the tidal pattern is accommodating, the mountain will be beautifully reflected. Even if the dawn is cloudy, it is worth walking along the river, because the sun often escapes momentarily from the grey curtain to give a dramatic light to the scene, creating a special mood for a few brief moments.

Whether or not the sea can be seen from Tidal River, there is always the dull boom of the surf rolling over the sands of Norman Bay. Day and night this sound is heard throughout the resort; not loud enough to intrude, yet there is an ever-present aware-ness, a sense of pleasure in knowing that the ocean is so close. Another persistent sound during daylight hours is the call of the Silver Gulls. They hover constantly over the resort, seeking free meals from the campers' food scraps. Like gulls everywhere, they fight and squabble among themselves for the titbits being thrown to them, their raucous, sometimes vicious, cries filling the air for as long as the food lasts. When no more is forthcoming, they quieten down and disperse, leaving a few hopefuls who wait around in case the camper has more to offer. There are always plenty of immature gulls. These young ones are nearly as big as their parents, and their high, feeble squeals sound quite incongruous coming from such large birds — especially when compared with the harsh tones of their elders. Although they generally breed on the nearby islands, some gulls are foolish enough to make their nests around the Tidal River camping ground, with disastrous results. If visiting children don't destroy the eggs, ravens or foxes will take them.

The gulls may be tiresome at times, but the Crimson Rosellas, which also abound, are not. These vividly coloured, rather comical birds are a constant source of delight, and the resort would be poorer without them. Unfortunately, they too enjoy being fed by visitors — unfortunate in that in a national park the birds and animals are supposed to be wild and free, not reliant on humans for their food. There is a very

A lichened granite boulder is reflected in the still water of Tidal River.

real danger the birds could stop feeding for themselves, and after peak holiday seasons, when few tourists are around, could die because they have come to rely on hand feeding — this has actually happened elsewhere in Australia. At Tidal River the rosellas are quick to find out who has the birdseed, and with chattering squawks they flutter in from all directions, presenting a flurry of brilliant crimson and blue feathers as they attempt to find a foothold on any part of the feeder's anatomy. Some of them are very tame, and this may be their downfall. In a city in far northern Australia, I met a man who had a backyard aviary full of budgerigars, among them one lonely-looking Crimson Rosella. He proudly admitted to me that the rosella had come from Wilsons Promontory. Not only is this very cruel, but also it is highly illegal and offenders can expect to pay huge fines, or face jail.

There are many other birds around Tidal River, although none in the same numbers as the gulls or rosellas. In quiet moments they come extraordinarily close to the camps. I have watched a group of Red-browed Finches busily feeding around my campervan, and on one occasion a perky little brown female wren took advantage of the open side door to hop up on to the step, then venture into the van to forage for crumbs on the floor. She didn't seem at all perturbed at my presence. The male Superb Wren, so beautifully marked with bright blue, is often seen around the camp, too. Like his brown mate, he is constantly on the move, as a result it is difficult to have a really good look at this endearing little bird. It is easier to watch the honeyeaters as they quietly feed off the spring blossoms, or the kookaburras who sit for long periods in the trees, displaying an air of utter concentration as they eye the ground for fat grubs. Because there are so many different kinds of habitat available for the avifauna, the Promontory has an astonishing variety of birds. Over the years more than 270 species have been sighted.

Tidal River near the camping ground, early morning

Silver Gulls attempt to grab a small morsel of food at the Tidal River camping ground. This common Australian seagull favours a wide variety of feeding grounds, ranging from beaches and coastal waters to city parks, inland paddocks and lakes, and rubbish dumps.

Crimson Rosella. In spring, when the tea-trees are in bloom, these rosellas like to feed on the nectar in the centre of the flower, and, in doing so, strip the blossoms from the branches. The immature Crimson Rosella is predominantly green, but irregular reddish patches give it a rather motley appearance.

A Kookaburra carefully scans the ground, looking for prey. This bird is the largest of Australia's ten species of kingfishers, which are noted for their powerful beaks.

Tidal River, from the footbridge. The early morning sun breaks through the grey curtain of cloud to create a special mood for a few brief moments.

Whale Rock

18

NORMAN BAY

Norman Bay's superb wide beach stretches for one-and-a-half kilometres between Little Oberon to the south, and Pillar Point to the north. As many walkers find, its distance is deceptive because the bold, high headlands at each end make it appear shorter than it is. Of all the Promontory's ocean beaches, this is the safest for swimming, despite the west coast's ever-present heavy swell that brings in the white-capped rollers to break over the flat sandy bottom beaches. And the water is cold — refreshingly cold during the hot summer days.

A path leads from the camping ground through the bush to the southern end of the beach and the Oberon Bay walking track, but it is much nicer to walk along the water's edge and be fanned by the fresh sea air. This is a perfect beach for walking or running: one cannot help but rejoice in the vast expanse of sand and sea. Because of its proximity to the resort it is a popular place, but no matter how many people are there, you can always find a quiet spot and feel that the beach is all yours.

Late afternoon, Norman Bay.

Norman Bay, from the Oberon Bay track.

Little Oberon, Norman
Bay.

South end of Norman
Beach.

TIDAL OVERLOOK

The Tidal Overlook track starts at the footbridge, from where the wide path trails easily up the hill. After about ten minutes' walk you will come to a clearing, opening on to a good vista of Norman Bay. From this vantage point the extent of the beach can be appreciated, especially at low tide, when a wide area of sand is left exposed. A little further on the track forks, and the right-hand section winds up the hill, leaving behind the tunnel-like tea-tree thickets and opening out to the heathland. As the path crosses the saddle, marvellous views unfold on both sides, and you are again conscious of the sound of the sea murmuring in the distance. There is an even better view at the Tidal Overlook lookout, where a huge boulder is situated at the end of a side track. (A short ladder helps you climb to the top.) From the boulder the path continues downhill to the main road. It is possible to return to Tidal River this way, making a round trip of nearly five-and-a-half kilometres. But the bitumen road with its traffic is not very appealing; and like me you may prefer to retrace your steps. I am always happy to do this, because the Tidal Overlook views are so good. Besides, there is always the possibility that you will discover different things on the return journey.

Tea-tree, by the Tidal Overlook track

Leonard Bay, from the Tidal Overlook track.

Norman Bay and Little Oberon, from the Tidal Overlook track.

PILLAR POINT TO SQUEAKY BEACH

It is hard to believe that Pillar Point, lying at the southern end of Norman Bay, was once covered in eucalypt forest and contained magnificent stands of Blue Gum.

In 1927 a bad fire swept through and destroyed the trees, and now an almost impenetrable scrub consisting of mainly tea-trees, she-oaks, and Kunzea covers the area. This is a classic example of the destruction a fire can wreak on natural wilderness.

The view from Pillar Point is worth seeing, and can be included in the easy walk to Squeaky Beach. The track starts from the footbridge at Tidal River, and shortly after the Tidal Overlook turn-off it forks, with one path going to Squeaky Beach and the other to Pillar Point. From the fork it is only about ten minutes' walk to the Point. The first view opens on to Squeaky Beach. Although the track gives the illusion that it ends among the rocks, you will have to go on further, first passing around an enormous boulder, and then out on to a natural platform of granite. From here there is a rewarding panorama over Norman Bay and the surrounding mountains. The first time I went to Pillar Point, I was surprised to find the vantage point so high — the track ascent had been deceptive. While I was there, a large group of children came on to Norman Beach, appearing like a mass of moving matchsticks as they ran along the sand. Even the waves looked minute as they chased each other to the shore.

It takes about thirty-five minutes from Pillar Point to reach the northern end of Squeaky Beach. The path zig-zags easily down the slope, giving lovely views over the beach and the huge boulders topping massive slabs of granite that slide into Leonard Bay, just below the path. If there is a strong swell, waves crash over the rocks, sending great fountains of spray into the air. Once out on the beach it is but a short walk to the boulders at the northern end. The pile lying on the sand marks the beginning of a long line of chunky rocks extending along the headland to Leonard Point and offering many hours of challenging scrambling. Occasionally, king waves sweep over the rocks closest to the water, but in spite of the risk involved, the rocks have a great attraction for foolhardy fishermen. Recently someone was washed off these rocks and drowned.

Most people visiting Squeaky Beach take their cars from Tidal River to the beach's car park, just off the main road. From the car park a few minutes' walk brings them out to the beach at the far northern end.

Leonard Bay

Lichened granite

Norman Bay, from
Pillar Point.

Squeaky Beach, from the southern end. Just before it reaches the beach, the walking track from Tidal River passes a collection of chunky tors topping slabs of granite that descend into Leonard Bay.

Squeaky Beach, the northern end. These rocks are part of the great pile of tors that lies on the beach, near the car park.

LILLY PILLY GULLY

This beautiful area provides one of the most popular short walks in the park, and the National Parks Service chose well in selecting it to be a nature walk. The wide track loops from the Lilly Pilly car park for five kilometres through superb bushland, ranging from heaths and eucalypt forest to lush fern gullies. Officially, it starts at the lower end of the car park, returning via the lower slopes of Mount Bishop. Many people find it much easier to start at the Mount Bishop end, as there is less uphill walking — an important consideration for the elderly. However, quite a few people prefer to walk only to the picnic ground and back — in this case, the easiest way is via the track starting at the lower end of the car park.

About half way down the trail, in the gully itself, a clearing has been made into a delightful picnic ground with an unobtrusive table and chairs. The self-elected custodian of the area is a large and extraordinarily tame Swamp Wallaby, who rarely fails to make an appearance in order to inspect the visitors and their food supply. He usually does very well, because few can resist those soft, imploring brown eyes. It was not lunch time when I first met him, but shortly after I arrived at the picnic ground, I heard a thumping noise behind me. I turned to see the wallaby heading purposefully towards my camera bag, which was on the ground. He peered inside inquisitively, then gave me a reproachful look, as if to say 'Where is the food?' I tried to coax him to pose for a photograph in the sunlight, but as I had no tasty morsel to offer, he refused to co-operate.

Another endearing but much less obtrusive mammal sometimes seen in the Lilly Pilly area is the Koala. These creatures are nocturnal and are not commonly active during the day — although walkers might be fortunate enough to see one moving around the trees very early in the morning. Generally they can be discerned only as furry grey balls curled up in the fork of eucalypt trees; should one be disturbed, it will probably peer down with a slightly vacant expression at the intruder for a minute or two before settling back to sleep. Overnight campers at Sealers Cove often hear the harsh call of the adult male, and if they are unaware that Koalas make such a noise, they may be more than a little startled when out of a nearby tree comes a sound remarkably like that of a hotted-up motor bike — or an angry wild pig.

The most beautiful and interesting part of Lilly Pilly Gully lies along the short detour path that loops from the picnic area through a section of temperate rainforest. This is one of the Prom's loveliest paths, and no nature lover hurries through here. After leaving the picnic area, you enter a green world of lush vegetation, dominated by fountain-like Soft Tree-ferns that form lovely canopies filtering sunlight through an exquisite tracery of a million tiny leaves. Fishbone Water Ferns spill along the stream running through the gully, and Swamp Gums, Blackwoods, and thickets of thin-stemmed paperbarks rise above the water and tree-ferns. Here and there lie the remains of huge trees, their rotting trunks now covered in thick green moss and decorated with discs of white fungus. In winter, the dark green Lilly Pilly trees are massed with white or purple berries.

Lilly Pilly Gully

The Loop track, Lilly
Pilly Gully.

Sometimes, where the path is close to the stream, you can see curious holes in the ground, measuring about four centimetres across. These are the homes of the Freshwater Yabby. Unlike the species that inhabits only dams and creeks, this yabby, which grows to about eight centimetres, lives in water at the bottom of burrows, and comes out at night to forage through leaf mould for worms and organic matter. It lives in the one home all its life, often extending the original burrow into a network of tunnels.

Before you leave the picnic area, sit down for a few moments and consciously *smell* the bush. If it is spring, the air will be full of the wonderfully sweet aromas released from flowers and vegetation by the warm sunshine; if it has just stopped raining, the scent will be fresh and tangy. Then listen to the concert of bird calls that fills the air — you may hear a Grey Thrush, an Eastern Whip-bird, an Olive or Golden Whistler, a Little Wattle-bird, or even the melodious Grey Butcher-bird. If you sit quietly enough, some of these may even show themselves: the Southern Yellow Robin is particularly friendly.

Swamp Wallaby.

Koala.

Yellow Stringy-barks
*(Eucalyptus
muellerana).* Walkers
pass through these
majestic, fibrous-barked
trees at the end of the
Loop track, at the picnic
ground.

In springtime new
fronds of the graceful
Soft Tree-fern unfold, at
first resembling ornate
walking sticks. The top
of the fern's trunk is a
natural compost heap,
collecting leaves, twigs,
seed pods, and other
debris which falls from
the upper storey of the
rainforest.

Look up to the top of a
Soft Tree-fern on a
sunny day, and see the
beautiful patterns
created by the fern
fronds. Soft Tree-ferns
flourish in Lilly Pilly
Gully's pocket of
temperate rainforest.

TONGUE POINT

In the days of the Darby River Chalet, the Tongue Point walk was extremely popular, but now that accommodation is based at Tidal River, people tend to ignore it. This is a great pity, because it is one of the best short walks in the park. It combines magnificent coastal scenery with a wide variety of vegetation, and in spring its range of wildflowers is superb.

There are two tracks to Tongue Point. One starts from the Darby Saddle, situated on the main road six-and-a-half kilometres from Tidal River; the other from the Darby River car park, five kilometres further on. Because both tracks are worth seeing, ideally you should start the walk at Darby Saddle, and return to the Darby River car park where a vehicle has been left. Of the two routes, the longer track from Darby Saddle is the more interesting, although it does have a fairly long, steeply-graded downhill section before it joins the easier Darby River track, which continues on to the Point. Before it meets the other path, the Saddle track passes through some exceptionally beautiful eucalypt forest, with groves of Drooping She-oaks, and a couple of good lookouts. Once you are out on the open heathland, there are constant coastal views for some time.

The track ends at a mass of granite boulders that falls sharply into the sea, then rises again to form the rocky end of the point. This landform has almost the appearance of an island, and in fact at certain times of very high tide it is cut off from the

Tea-trees, Tongue Point track. Many of the Promontory's walking tracks pass through similar dense, tunnel-like thickets.

Here huge granite
boulders, many of them
covered in brightly-
coloured orange lichen,
lie tumbled over the
headland.

At Tongue Point

Hippopotamus Rock,
Tongue Point.

rest of the headland. The huge, angular boulders provide good opportunities for exploring, and sheltered spots invite you to rest and contemplate the heavy swell in its rhythmic rise and fall against the gaily-coloured lichened granite. In one place near the Point the track passes by Hippopotamus Rock, a granite headland that provides a good spectacle as it tumbles dramatically into the sea. In a sheltered nook below the path you can spend some pleasant moments watching the waves foaming over the rocks, and listening to the hollow booming sound of the swell as it pounds into unseen caverns below.

The grades are much easier on the Darby River track, which is less than five kilometres walk back to the car park. At one point a side path leads down to Fairy Cove, where there are more groves of beautiful She-oaks. Not far from the car park, the bushes lining the path are low enough to permit a view of the sheer coastal slope running northwards. One remarkable feature to be seen here is the demarcation line between the two types of terrain, granite and sand. The line shows quite distinctly where the grey granite gives way to red-coloured sandstone.

Drooping She-oaks
(Casuarina stricta)
misted with rain, on the
path to Sparkes
Lookout.

Sparkes Lookout

This lookout point is about three-quarters of an hour's walk from the Darby Saddle car park, and five minutes from the Tongue Point track.

DARBY BEACH

Darby Beach is quite different from the other beaches around Wilsons Promontory. Except for some granite at the end of the southern headland, there are no boulders studding the sands; instead, a line of colourful sandstone cliffs stretches for about 500 metres towards Waratah Bay before subsiding into low dunes. Fretted by the elements into fascinating whirls, ridges, and crenellations, these superb cliffs are at their best in the late afternoon, when the low sun flares the sandstone to a deep, warm gold. Providing there is not an extreme high tide, there is good walking past the cliffs. Access to the beach is from the Darby River car park, a spacious, grassy clearing lying by the river, eleven-and-a-half kilometres from Tidal River. The fifteen-minute-long track is wide and level, although it passes through very loose sand before reaching the beach. The cliffs lie just beyond the Darby River, which cuts a shallow course through the sand before running into the sea.

If you want to see a wombat, Darby River is one of the best places to find one. These animals like to graze the grass covering the dunes that lie behind the beach, and often roam along the beach end of the Darby River at dusk. If you move quietly, it is possible to get quite close to one. The creature may stop eating and stand warily waiting to see what your next move will be; but if you don't seem very threatening, it will probably continue to nibble at the grass, and not object too much if you wish to take a photograph.

Wombat, Darby River. It is not unusual to see these animals around the beaches, grazing the dunes' sparse grassy covering, and sometimes you can get quite close before they scamper off.

Sandstone cliffs, late afternoon, Darby Beach. Worn by the elements, these cliffs take on a warm glow when the sun is low in the sky.

WHISKY BAY AND PICNIC BAY

Both Whisky Bay and Picnic Bay have car parks from which short walking tracks lead to their beaches. Picnic Bay is tucked between Leonard and Picnic points, and like many of the Prom's beaches, is wide, with sands sweeping to a rocky point. At the northern end, a five-minute walking track goes over the hill of Picnic Point to Whisky Bay's beach. A great favourite with visitors, tiny Whisky Bay is full of character. Much of its charm is in its size — only about 260 metres of sand lie between the huge piles of boulders that dominate each end.

One cannot help thinking that there must be a story behind a name like Whisky Bay: in fact, there are two. One has it that the bay was named for the colour of the water in the creek, which resembles weak whisky — but then most of the Promontory's watercourses at low levels are of an identical colour, the results of the tannin present in decomposing tea-trees. The other version, which is rather more imaginative, claims that many years ago a crate of whisky fell off a vessel and was washed on to the beach, and that the bay was named after this incident.

Actually, from time to time some very strange things do get washed on to the shores of the Prom — not so long ago a container full of furniture landed on one of the western beaches.

Whisky Creek. Dunes covered in sparse vegetation lie back from the beach at Whisky Bay, and the creek trickles down from the hillside to flow beside them, in places spreading into small pools before it enters the sea.

Whisky Bay, from Picnic Point. At the end of the point there are some excellent views over Whisky and Picnic bays and the great heaps of boulders strewn along this small headland.

MOUNT BISHOP

Of all the views from high vantage points on the western side of the Prom, I find the one from Mount Bishop the most appealing. The hill is high enough to give a magnificent perspective, yet low enough for the observer to discern detail. By contrast, the more popular Mount Oberon summit presents a more spectacular panorama, but lacks Mount Bishop's soft beauty.

The Mount Bishop track starts from the main Lilly Pilly Gully path, where a sign advises that the last two kilometres are rough; but most walkers find that it is just as good as the tracks on the Prom's eastern side. The grades are easy — easier than those on Mount Oberon. From the car park the walk is four kilometres long, and the average walker should have no trouble in reaching the summit within the hour.

The track passes through some lovely vegetation, and once it has reached the top, it branches off to various lookout points among the huge boulders scattered over the peak. Looking westwards from the top, the view extends over Leonard and Tongue points, and Norman Island. The sea shines like metal in the afternoon sunlight; and if the sky is full of clouds, they often seem so low that you feel you could reach out and touch them. From one spot, you may find yourself peering cautiously over the side to see below the giant cliff-like slab of granite that dominates the peak. The best vantage point among the boulders lies at the end of the path that leads to the extreme left and overlooks Tidal River and Norman Bay.

Common Correa
(Correa reflexa)
Flowering time: spring.

(right)
Tea-tree, Mount Bishop
Flowering time: spring.

The view over Norman Bay and Tidal River, from the summit of Mount Bishop. This boulder-studded peak is a dominant feature in the views from Squeaky Beach and the Tidal River Road.

Mount Bishop, from the Squeaky Beach car park.

MOUNT OBERON

I first walked to the summit of Mount Oberon on a wild and stormy autumn evening, when the sky was full of clouds pouring in from the south, and the wind howling miserably through the scrubby vegetation. At first I could see nothing, because the clouds enveloped everything on the mountain; but suddenly they lifted to reveal a stunning view of headlands, bays, high granite ridges, and timbered valleys lying like a great map below. Periodically, heavy curtains of grey rain drifted through the Tidal River valley, and then a squall would briefly dash itself against Mount Oberon. On one occasion the sun, now low on the western horizon, managed to break weakly through to streak the sea with silver, and to highlight the clouds. The scene was breathtaking.

The track starts from the Mount Oberon car park at the Telegraph Saddle, and follows the service road that winds up the eastern side of the mountain to Telecom's microwave relay station. From there a footpath leads up the final granite section to the peak. The grades of the road vary from easy to moderately steep, and you should allow an hour for the journey up. Most people return in about three-quarters of an hour. The road passes through thickets of banksias and other tall heathland plants, lovely eucalypt forests with Messmate and stringybarks predominating, and around damp gullies where ferns and mosses thrive.

This is a popular walk, and many people like to take it in the late afternoon in order to see the sunset, which can be superb — weather permitting. If this is your intention, you will probably find that by the time you are on the last part of the return trip, night has fallen. It is a good idea to take a strong torch — not so much to light the road, which is clear enough in the night light, but to see nocturnal creatures which could be in the trees. There are three species of possum common on the Promontory — the Brushtail, the Ringtail, and the Pygmy Possum, and you may be lucky enough to spot them.

Birdlife at the summit of Mount Oberon can be surprisingly varied. One might see Tree Martins, swallows, hawks, and White-breasted Eagles, to name just a few. Once, while I was waiting for the sun to set, two Southern Yellow Robins came and kept me company, showing a remarkable degree of tameness. Nature also has its less pleasant side, however. If you hope to escape those summer pests, the marsh flies, you will be disappointed. On still days, the sun-warmed air is filled with their buzzing drone.

Mount Oberon, from
the Tidal River Road.

From the summit of
Mount Oberon, late
afternoon.

49

SEALERS COVE

Of all the long walks in the park, Sealers Cove is probably the general favourite. An ideal day trip, it takes two to three hours each way from the Mount Oberon car park for the average walker. Most of the nine-and-a-half kilometre route is surprisingly well graded, only a few spots being noticeably steep. It is a beautiful track, passing through shady eucalypt woodlands, tunnels of tea-tree thickets, moist rainforest, groves of superb tree-ferns, and swamplands that can now be enjoyed from boardwalks.

If I am not camping overnight at Sealers Cove, I like to start at first light in order to be at the cove as early as possible. One summer morning, I remember stepping out on to the beach to find the brilliant blue water lying like a still pond, the soft morning light warmly touching the wooded hills that surround the cove. There was a deep sense of peace. Not a soul was in sight, and the golden sand lay smooth and undisturbed, all footprints erased by the previous night's high tide. At the southern end, Sealers Creek mirrored the nearby vegetation and two herons and a cormorant feeding in the shallows. A couple of hours later a wind had chased away the reflections, the birds had gone, and small waves ruffled the sea; several large and noisy groups of people swarmed over the beach. The peace had gone.

Moss-covered rocks,
below Windy Saddle

Sealers Cove track.

The camping ground lies just beyond the creek. Usually it is easy enough to cross at low tide. Once, when two rangers carrying backpacks had to cross during a king tide, the water came up to their chests. This is a popular camp, because Sealers Cove is so beautiful most people want to take the time to see it properly. Nights here are marred only by the occasional raids of the rogue wombat, Lester. This particular wombat has been at the Sealers Cove camp for years. I remember him from as far back as 1975, when he attempted to feed from our supplies. Once the rangers tried to move the truant to another part of the Prom. He was caught without too much difficulty, but broke out of his wooden crate before he had left the cove.

The Sealers Cove track is part of the magnificent circuit trail which links Sealers and Refuge coves and Waterloo Bay to the lighthouse track, ending at the car park. Most people take two to three days to complete the walk, camping at the outstation camping grounds, but there are a few hikers who complete the thirty-seven-kilometre round trip in one day.

Late afternoon at Sealers Cove. The tide is in, filling Sealers Creek, which usually cuts a path through the sand near this old dead tree.

Sealers Cove, from the
boulders lookout by the
Refuge Cove track.

An old melaleuca
bends over the sand at
Sealers Cove, near the
start of the Refuge Cove
track.

Crested Terns, Sealers
Cove.

REFUGE COVE

A hiker's fitness is truly tested on the first part of the Refuge Cove track. Starting near the camping ground at Sealers Creek, the path leads up an extremely steep hill for about the first half-an-hour, then eases somewhat into an average uphill grade for nearly two kilometres before the descent. The distance between the two coves is a very long five kilometres. A highlight on the steep section is a lookout point from some huge granite boulders lying on the right side of the path, about twenty minutes' out of Sealers Cove. The view over the cove is one of the Prom's great panoramas, and if you are going on a day walk to this area, it is worth going up this steep hill in order to see it.

Refuge Cove actually consists of two small coves that share a single opening to the ocean, approximately 200 metres wide. As its name implies, it provides the safest anchorage along the Promontory's east coast, and many boats have sought protection here from the storms of Bass Strait. Over the years it has been a popular place with boatmen, many of whom have left reminders of their visit by writing names on the boulders heaped along the shore. I was horrified when I first saw Refuge Cove some years ago. Under a grey sky, the boldly printed words that covered the rocks shrieked their vulgarity, showing up against the muted colours as the dominant feature of the shoreline. However, when the sun shone the next morning, I found the beauty of the place quite overwhelmed man's crude efforts, and the vandalised rocks, now in shadow, no longer had the same impact. The tide was high; and small, gently foaming waves lapped almost to the roots of the gnarled Coast Banksias leaning over the sand of the small beach. Despite the sounds of a colony of gulls squabbling among themselves along the water's edge, the scene was tranquil.

Today Refuge Cove has been returned to its former beauty, thanks to a devoted group of people called 'Friends of the Prom' who, in 1981, organised the first of several work parties to remove the names from the rocks. Wilsons Promontory is fortunate to have this caring group. Among other activities, they have removed glass from beaches and carried out certain weed control measures — projects the rangers have neither the manpower nor the money to undertake.

Pleasure craft continue to call into Refuge Cove, and for hikers there is a camping ground in a beautiful setting by Cove Creek. Nearby is the start of the Waterloo Bay track which trails up a long hill to Kersops Peak, a walk that takes about an hour. At the peak there is a well-placed group of boulders, which make an ideal resting spot and vantage point for the magnificent views that stretch over Waterloo Bay and the lighthouse on South East Point.

From Kersops Peak.

Refuge Cove.

WATERLOO BAY

The best scenery of the circuit lies between Refuge Cove and Waterloo Bay. After the long uphill walk out of Refuge Cove to Kersops Peak, and the easy descent overlooking the splendid coastline, the path crosses two beautiful beaches in North Waterloo Bay. At the far northern end, just below the walking path, piles of boulders shelter an expanse of sea to form a miniature cove. On a warm summer's day few hikers can resist cooling off in the crystal clear turquoise waters of this romantic little cove. From here the track, although shaded, becomes extremely rough. For most of the distance it hugs the rock-strewn coast, passing lichen-capped boulders and great tors that jumble with grey, finger-like stones pushing into the bright blue sea. Little Waterloo Bay, with its dazzling white sand and surrounding hills, is superb. Behind its dunes is the camping ground, sheltered in a cool forest.

Twenty minutes away from the camping ground is Waterloo Bay, where more white sands stretch to the distant but prominent headland. The sea in this bay can be treacherous for swimming — as I learned the hard way. Thinking it would be safe just to splash around at the water's edge, I was thoroughly alarmed to find a strong undertow dragging me towards the open sea. I had considerable difficulty in getting out.

The path from Waterloo Bay to the lighthouse track is easy, and takes about one-and-a-quarter hours. The swamps lying behind the bay are now spanned by a series of boardwalks, which enable walkers to enjoy, rather than curse, this fascinating area. Who could forget the days when a maze of vague tracks meandered through a jungle of paperbarks, sword grass, and soft black mud which, in very wet years, was more than a metre deep in places? I first walked through these swamps in a summer following a wet spring, on a day when the mercury stood at 38°C. The mud was frightful; our backpacks kept tangling

in the paperbarks; and when I clutched a clump of sword grass in order to regain by balance after floundering in the mud, it was like grabbing a hundred carving knives. Confused footprints went in all directions, and as it was impossible to keep a straight course, we had to be careful that we didn't go round in circles.

About half way between the swamps and the lighthouse track, the path passes by a spectacular group of tors standing on a high ridge. In years past they were called the Mussolini Rocks, after the Fascist dictator. According to the imagination, the tors wear many faces, some obscure, others more distinct; from a distance they resemble a strange collection of chess pieces.

Swamp, Waterloo Bay track.

Waterloo Bay.

Tors, Waterloo Bay track.

OBERON BAY

The wide, sandy beach at Oberon Bay, on the west coast, stretches for over two kilometres between its guardian granite headlands. Mount Wilson, one of the higher peaks of the Promontory's rocky spine, forms a superb backdrop to the bay — particularly when morning mists wreath the summit. The six-kilometre walking track from Tidal River comes out to the bay at the northern end, by Growlers Creek. This easy walk, which takes you across Norman and Little Oberon bays, is very scenic; and in summer many hikers stay overnight at the camping ground set beside Oberon Bay's other freshwater stream, Frasers Creek.

Oberon Bay is linked to the lighthouse track by a path running through the large sand-drift immediately behind the beach.

Red markers indicate the route through the dunes. At times the walking is heavy going through the loose sand, but the beauty you will find is worth the extra effort. In the early mornings and late afternoons, when the sun partially veils them in shadows, the dunes are magnificent. The low light also causes the ripples covering the sand to stand out so boldly that the ground creates the illusion of flowing like water. The exact cause of this sand 'blowout' is not known but probably the dunes, although they lie in a very exposed part of the Promontory, were stable before the coming of Europeans. Many people believe that the grazing of cattle in this area during the last century triggered the drifting sand.

Growlers Creek,
Oberon Bay.

Sand dunes behind Oberon Bay, with Mount Wilson in the distance. Although the dunes appear to be creeping eastwards, they have in fact started to stablise.

THE LIGHTHOUSE

Although quite a few very fit people happily complete the thirty-seven-kilometre round trip to the lighthouse in one day, taking five to six hours each way, most hikers prefer to walk at a more leisurely pace and stay overnight either at Half Way Hut or at the Roaring Meg camping ground. Passing through a variety of interesting vegetation, the track starts at the Mount Oberon car park and follows the fire access road for most of the way. At Martins Hill a signposted detour cuts through the bush to the picturesque Roaring Meg camping ground, and is picked up again at the lower campsite by Roaring Meg Creek. Nearly three kilometres later it briefly joins the road until another bush path branches off to the left, crosses the saddle below South Peak, and leads down a fairly steep hill to the lighthouse gates.

I am always astonished at the number of people who take this popular long walk without first reading the national park's walking notes, or the big notice at Tidal River which advises that the lighthouse is open for inspection only on Tuesdays and Thursdays, and then by appointment only (the lighthouse is *not* part of the park). Many walkers arrive at the lighthouse gates and find them locked; and just as many, not knowing that the final bush path to the lighthouse exists, miss the signpost and continue along the road for a few kilometres before it peters out into the forest. For the same reason, some even miss the short cut from Martins Hill, thus adding more kilometres to an already long walk. On a hot day, this could be serious.

Once at the gates, the track runs steeply up the hill to the settlement and the lovely old stone lighthouse which stands, exposed to keen winds and stormy gales, at the edge of South East Point. Contrary to popular belief, this sheer, high granite headland thrusting into Bass Strait, is not the most southerly tip of the Promontory: this distinction goes to South Point, lying just to the west.

From the lighthouse grounds there are some breathtaking views of the coast, its beauty enhanced by dramatic foregrounds of the massive granite boulders that are such a distinct feature of South East Point. Many of the rocks have been wonderfully shaped by the elements. Of these, the most extraordinary are Skull Rocks, a collection of huge cavernous tors that lies inside the lighthouse reserve. Walk among these monoliths and see at close quarters the full extent of the weird hollows, caves, and indentations. It is an awesome experience.

Grass Tree

The lighthouse, South
East Point. Built of
Melbourne basalt and
local granite rubble,
and opened for service
in 1859, this is one of
Australia's major light
stations. It is managed
by two lightkeepers. As
there is no vehicular
track into the lighthouse
reserve, the station is
serviced by helicopter.

Rodondo Island, from
the lighthouse reserve.
The first recorded
landing on Rodondo
was made in 1947 by
John Béchervaise and a
team of Geelong
College school
students.

Coastal view from the
lighthouse reserve.

Skull Rocks. These spectacular tors dwarf the vehicle standing nearby. The view shown here is from the track leading down to the lighthouse gates.

In the 1970s the beautiful crystal glass prisms, which for years had effectively and economically beamed the light from a single globe, were dismantled. They were replaced by four vertical metal panels covered with a total of forty-eight powerful lamps. The prisms are now in the Port Albert Maritime Museum.

THE PROM'S ISLANDS

Sixteen offshore islands and numerous rocky outcrops form part of the Wilsons Promontory National Park. Surprisingly enough, the conical-shaped Rodondo Island, which can be seen from Mount Oberon and many other points in the south, belongs to Tasmania. The most familiar islands are the double-humped Norman Island, off Leonard Point, and the four in the Glennie Group, seen from Norman and Oberon bays. Also visible from Norman Bay is the Anser Group, which lies south of the Glennies; of this group, the most distinct is Cleft Island — or Skull Rock, as it is more popularly known. The north face of this squarish granite island is obscured by what looks like a dark shadow, but which is actually a gigantic cave. At least one recorded ascent to the cave has been made. After a five-hour climb using ropes, the research team found a cannon ball, and the bones of birds never before known in these waters.

Landing on most of the islands is extremely difficult because of their rocky and often sheer slopes, and the treacherous swells of Bass Strait. The light beacon on Citadel, one of the Glennie Group, is now serviced by a helicopter; previously a gantry and a crude flying fox hauled the equipment to the top. On Great Glennie, the only spot suitable for landing a boat — and then only at low tide and in calm seas — is on a tiny beach in the comparative shelter of the island's steep, rocky slopes.

Access to the islands is restricted. Nearly all are important breeding grounds for a wide range of sea birds and visitors must gain permission before attempting to land on them. On Kanowna Island, in the Anser Group, there is a large colony of Fur Seals. In places, their sleeping bodies cover the granite slopes in hundreds. When a boat draws near the slopes erupt into bleating, barking life, and it's an astonishing sight to see the seals sliding en masse into the water, as if the rock face is literally tilting them into it.

Cleft Island (Skull Rock).

Fur Seals, Kanowna Island.

THE NORTHERN END

In the past, like many who visit Wilsons Promontory, I never really considered walking in the northern section. The south had so much to offer, both in diversity of tracks and wonderful scenery, there never seemed enough time to discover even this part satisfactorily. Until recently, there was very little information or encouragement for those who wanted to walk in the north. At a glance the distances seemed formidable, and for some odd reason I didn't think there would be much to see. How wrong I was! Not only is there excellent walking but also some wild and beautiful scenery. Taking advantage of the various base camps available, the distances to Johnny Souey Cove and Five Mile Beach compare favourably with some of the southern tracks, and for those who don't want an overnight camp, there is a short walk to Millers Landing on Corner Inlet and a pleasant day walk to Chinamans Creek, situated on the Five Mile Road.

All walks start from the Five Mile Gate car park, situated a few kilometres off the main Tidal River Road. The left track, which leads to Millers Landing, has been made into an interesting nature walk, complete with a brochure guide. It is an easy walk over a wide track that passes through splendid forests of Saw Banksia (one of the most spectacular species of banksia) before opening out to grand views over Corner Inlet. It takes about three-quarters of an hour to reach the small beach at the foot of the old landing steps. If the tide is out, exposing mud flats that are firm enough to walk over, the area is much easier to explore. This is also the best time to see crabs scuttling over the flats, other interesting forms of marine life, and the exposed tangled root systems of the mangroves which have colonised the inlet. The mangroves, the most southerly in the world, give the inlet its unique character, in strong contrast with the Prom's other bays and inlets.

On one occasion I reached the small beach at the end of the track only to hear murmuring voices coming from around the corner. I was surprised, for I knew there were no walkers ahead of me. Investigating, I soon discovered the 'voices' came from a large flock of black swans feeding in a nearby shallow lagoon. This memorable sight was very brief, as the birds suddenly took to the air in a flurry of feathers, and swiftly moved over the water as a great black cloud, their cries ringing around the inlet.

The main walking track through the north follows a rough fire access track known as the Five Mile Road. The most popular places off this track are Johnny Souey Cove and the Five Mile Beach. Fewer people continue on to Tin Mine Cove, which is a further seventeen kilometres from the Johnny Souey Cove turn-off. The only suitable day walk here is to Chinamans Creek, ten-and-a-half kilometres from the Five Mile Gate. With good views over Corner Inlet, the track crosses the lower section of the Vereker Range, passing through heathland and light woodland which suddenly and unpredictably changes near Chinamans Creek to a very beautiful and inviting forest, full of tall Messmates, Lilly Pilly trees, and the lush vegetation associated with the Prom's moist gullies. It is a very restful place, with an atmosphere that entices you to do nothing except absorb the beauty of the vegetation.

The same sort of peace can be found at Johnny Souey Cove, situated nearly twenty kilometres from the Five Mile Gate. The cove's small beach curves into low hills that protect it from the ocean's strong swells. The sheltered camping ground is at the scenic southern end, where granite rocks dot the sand, and a freshwater pool lies tucked into a corner of lush vegetation.

Who was Johnny Souey? Nobody really knows, but it is presumed he was one of the Chinese connected with a nineteenth century fish drying enterprise. A permanent reminder of their presence in the area is left in various place names such as

Mangroves, Millers Landing. This was the place where early visitors to the Prom left the boat to walk to the Darby Chalet.

Chinamans Creek, Chinamans Knob, and Chinamans Long Beach. The track in and out of Johnny Souey Cove passes over a high, steep hill which gives magnificent views over the coast and mountains — one of the best panoramas in the Prom.

There is more beauty at Five Mile Beach, which lies two kilometres off the main track. In a distinct atmosphere of wilderness, the beach of strikingly yellow sand sweeps for seven-and-a-half kilometres towards the high Cathedral Range; the other side of the Cathedral's headland is Sealers Cove. Later I commented to a ranger that the beach couldn't possibly stretch for that distance to the Cathedral — it looked barely three or four kilometres. 'Try walking it!' was his only comment. This beach also has a camping ground, situated by a freshwater spring at the far northern end. Another campsite used by walkers while exploring this area is at Barrys Creek, seven kilometres from the Five Mile Gate car park.

In the hot summer months, at times of high fire danger, these walking tracks are sometimes temporarily closed. They may also be closed when the freshwater streams at the camp sites dry up, as can happen in dry seasons. The Five Mile Road affords little shade, and dehydration can be a major problem for hikers during hot weather. The best time to walk in the north is during the spring, when the weather is cooler and the beauty of the heathlands is enhanced by a superb display of colourful wildflowers.

Chinamans Creek, on the Five Mile Road. Tall Messmates rise through lush vegetation in this beautiful gully situated ten-and-a-half kilometres from the Five Mile Gate car park.

Grey Kangaroos, with
the Vereker Range
behind.

Johnny Souey Cove, on
the Promontory's north-
eastern coast. At the
southern end of the
beach a small stream
cuts its way through the
sand before entering
the sea. This scene is
near the camping
ground, which lies
close to a freshwater
pool.

Rain sweeps across the view from the hill behind Johnny Souey Cove. This photograph was taken from the walking track looking over Five Mile Beach and the ranges beyond. On a sunny day the golden sands of this beach can be seen clearly from here, sweeping along the coast like a long, bright ribbon.

Five Mile Beach, with
the prominent
headland of Cathedral
Range in the distance;
around the corner is
Sealers Cove.

THE WILDFLOWERS

Few areas of similar size harbour such an abundance and diversity of plants as Wilsons Promontory. There are well over 700 species of indigenous flowering plants, ferns, and mosses — nearly a quarter of Victoria's total flora. The much-loved orchid family is well represented: over eighty kinds (almost half of the State's species) have been recorded. In addition to the indigenous flora over 100 alien plants have become established. Fortunately, only a few of these are troublesome. As the Promontory was once linked to Tasmania, it is not surprising to find that many of the species are also found in the island State. Some do not grow elsewhere in Victoria, while others grow only in selected spots.

Naturally, spring is the best time to see the wildflowers of the Prom. From October through to November, the heathlands, especially those in the northern part of the park, are full of colour. The tracks to Sealers Cove and Tongue Point are a botanist's delight; and when the masses of tea-trees are in flower, the coastal slopes look as if they are powdered with snow. Most spectacular of all is the display of showy White Kunzea, which grows along many of the walking tracks and around the car parks. The strong honey-sweet scent from its sprays of creamy blossoms so fills the air that it has become a distinct feature of a Promontory spring.

White Kunzea (*Kunzea ambigua*), Tidal River Road. Flowering time: Oct-Nov.

Twining Fringe-lily
(*Thysanotus patersonii*)
Flowering time: Sept-
Nov.

Pink Fingers *(Caladenia
carnea)*
Flowering time: Sept-
Nov.

Horned Orchid
(*Orthoceras strictum*)
Flowering time: Oct-Jan

Saw Banksia *(Banksia
serrata)*
Flowering time: Dec-
Feb.

79

Pigface
(*Mesembryanthemum*
sp.) North Waterloo
Bay.
Flowering time: spring
to summer.